Thumping on Trees

Richard D. Alexander

Illustrations by John Megahan

Richard D. Alexander

John Megahan

WOODLANE FARM BOOKS
Manchester, Michigan

Copyright © 2009 Richard D. Alexander

All rights reserved. No part of this publication may be reproduced or transmitted in any form or by any means, electronic or mechanical, including photocopy, recording, or any information storage and retrieval system, without permission in writing from the publisher.

ISBN: 978-0-9712314-4-3

Woodlane Farm Books
10731 Bethel Church Road
Manchester, Michigan 48158

www.woodlanefarm.com

Manufactured in the United States

Thumping on Trees

All the time the boy was growing up, and even across his entire life, he kept remembering that thumping on trees had helped him learn how to discover new and wonderful things that a person can appreciate, and remember forever.

Acknowledgments

I thank John Megahan for his superb illustrations, and his tolerance in satisfying me that the illustrations reflect accurately the circumstances of every event in the stories.

Lorraine Kearnes Alexander, my incredibly patient wife, and my soul mate for more than 62 years, has, as usual, helped and encouraged me at every step.

Most of all, for this book, I am indebted to Lorrie's and my two schoolteacher daughters, Susan and Nancy, and our four grandchildren: Morgan, Winona, Alex, and Lydia. As youngsters, these six never stopped insisting that I keep on telling them stories from my childhood in Sangamon Township—over and over and over.

Introduction

The stories in this book are true. They are told here, exactly as they are remembered, from two-thirds of a century ago, along the Sangamon River in Piatt County, Illinois.

When the boy in the stories grew up, he became a father, then a grandfather. He told the stories first to his children, then, later, to his grandchildren.

Henry, the name used for the boy in the stories, is not the real name of the boy who did all those things so long ago. It's a name the boy's father sometimes used for him, as a good-natured tease, when the two of them were working together somewhere on the family's farm.

The boy liked it when his father called him Henry. He would usually see a little smile on his father's face. He knew that using the name, Henry, was one of his father's ways of expressing some extra-special affection for him.

So, when the boy grew up, and became a father and a grandfather himself, he called the boy in his stories, Henry.

Using the name Henry in his stories helped the grandfather remember his father's heart-warming good humor.

Chapter 1

Thumping on Trees

Henry's favorite place for exploring was along a creek, and into the woods that grew along the river near his family's farm. He liked walking in the woods because he saw raccoons and possums and skunks and foxes and frogs and snakes and crickets and beetles and butterflies and owls and hawks and crows and red-headed woodpeckers. Sometimes he would see giant catfish or carp swimming in the river, little groups of minnows or sunfish darting about in the creek, or a pair of muskrats paddling across a pond. Sometimes he found Indian arrowheads, pretty feathers, brightly colored leaves, and other nice things that he kept for a while—or forever.

Henry always seemed to find something new in the woods. Once he found four baby raccoons in a hollow sycamore tree. Another time he found a fox den where there was a litter of five little foxes he could watch, as long as he remained hidden and was very quiet. He always had a good time looking for special things in the woods.

Henry had to walk a long ways to get to the woods. He had to walk down from the low hill where his family's house and barn had been built long ago, and then along the creek that wound through Walter White's pasture where the big white-faced cows were eating grass. He had to follow the creek to where it finally spilled its water into the big river. Sometimes he saw wild mallard ducks swimming in the creek and stopped to watch them to see if he could find out where they had hidden their nests.

One time Henry was walking along the creek through Walter White's pasture, past the willow trees and the oak trees and the tall skinny elm trees. In front of him a big reddish-brown squirrel flipped its tail, ran along a fallen log, and hurried up a tall skinny elm tree. Henry walked up to the tall skinny tree and looked for the squirrel. But the squirrel was on the other side of the tree, so Henry walked around the tree to look at it. The squirrel hurried around to the other side, away from Henry, and it also went up the tree just a little bit. Henry walked around the tree again and the squirrel ran around to the other side. The squirrel also went up the tree just a little bit more. Henry thought this was fun, so he began to run around and around the tree, faster and faster. The squirrel also went around and around, faster and faster, and it also went up, up, and up. It went up until it got right to the very tip top of the tree. Then it stopped because there was no more tree for it to climb. But Henry didn't stop. He ran around the tree one more time. The squirrel tried to run around the tree one more time too, but when it went up the tree a little bit, as it always did, there wasn't any more tree.

So the squirrel just leaped right out of the top of the tree and sailed down through the air with its legs all spread out. Down, down, down it fell, and it hit the ground with a big loud "plop" right in front of Henry.

Henry was very surprised, and he thought the squirrel must be hurt from dropping so far to the ground. But the squirrel was not hurt. It jumped up and ran away very fast to a big tree, flipped its tail, and hid where Henry couldn't see it any more. Henry laughed and laughed at the squirrel that had leaped out of the top of the tree. Then he went on down the creek toward the river.

When Henry walked down to the woods along the creek he sometimes stopped by a little grove of sassafras trees and cut himself a walking stick. Sassafras wood has green bark and a nice smell. It is very tough and strong. Henry's walking sticks were usually knobby. Henry liked to carry a walking stick that was knobby and felt good and smelled good. A walking stick was a fine thing for thumping trees. Henry liked to thump trees with a sassafras walking stick.

Henry thumped trees with his walking stick because he wanted to find a hollow tree. He knew that if he thumped a tree that was not hollow it would go "Thud! Thud! Thud!" the way a ripe watermelon sounded when Henry thumped it with his finger in his watermelon patch or at Emerson Evans's roadside watermelon stand: "Thud! Thud! Thud!"

But Henry knew that a hollow tree didn't go, "Thud! Thud! Thud!" when it was thumped. A hollow tree sounded like a drum. When Henry thumped a hollow tree with his sassafras walking stick, it would go "Bohnk! Bohnk! Bohnk!" like a watermelon that wasn't ripe when Henry thumped it with his finger. Henry understood that thumping on watermelons, or on trees, was trying to find an answer to an important question.

It was important to know if a watermelon was ripe or not, because choosing a watermelon that wasn't ripe was a big mistake. You had to carry the melon home, and then throw it away because it wasn't ready to eat. If you accidentally picked a green or unripe melon in the watermelon patch, it would be wasted because the thumper hadn't understood the sound of his own thumping. People who didn't know how to thump a melon to find out if it was ripe would ask Emerson Evans to thump the melon for them. Emerson Evans could always tell if a watermelon was ripe by thumping on it.

Sometimes people would ask him to plug the melon before they bought it. Mr Evans would then take out his pocket knife and carefully cut a small triangle in the side of the melon. He would make the sides of the triangle slant a little toward one other as they went down into the melon, so the piece he cut out would narrow down to a tip about as long as the knife blade. Then Mr. Evans would use the tip of his knife blade to lift the plug out of the watermelon. He would hold the plug out on the knife blade and let the customer bite off the ripe red part to test the quality and ripeness of the melon. If the melon didn't taste good and smell good, the customer would ask for another, and Emerson Evans would feed the bad melon to his neighbor's pigs.

Henry had learned from his parents about a great hollow tree across the river. It was so big that a man lived inside it. A man lived inside a hollow tree! Henry had never seen the man, but others had. And he had once seen the ashes of the man's fire, just outside the door to the great hollow tree where the man lived. After seeing the home in the hollow tree, Henry decided he would like to find his own great hollow tree, big enough that he could climb inside and make a special little forest hideaway for himself. He thought the inside of a hollow tree would be a good secret place in the woods. He liked thinking about such a tree, and about climbing inside it and keeping it a secret. That is why he regarded thumping on trees as a way of answering an important question.

Whenever Henry finally got to the woods, after walking through Walter White's long creek pasture, he always started thumping every big tree with his sassafras walking stick. But all the trees he thumped just said, "Thud! Thud! Thud!" Not one said, "Bohnk! Bohnk! Bohnk!" Not one of them was hollow. But Henry was persistent. He kept right on thumping trees anyway.

Then one day, when Henry thumped a certain big tree he had never seen before, it didn't go "Thud! Thud! Thud!" Instead it went "Bohnk! Bohnk! Bohnk!" Henry stopped and stared wide-eyed at the tree. He was really surprised that it went "Bohnk! Bohnk! Bohnk!"

But then, before he could say, "Jack Rabbit's Habit," there was a loud noise inside the big tree. And the noise went, "Pshflelshcrshlopflschlickl!" And then it went "Ploosh! Ploosh! Ploosh! Ploosh!" right up the trunk of the tree, higher and higher.

Henry's eyes were wide-open, and they went up the tree too, following the sound as it went up the tree. And when the ploosh-ploosh-ploosh-plooshes got near the top of the tree something suddenly flopped out of a big hole high up in the hollow part of the tree. It sat on the edge of the hole, and looked straight down at Henry with big round stern yellow eyes.

The "something" was a huge owl with feathers on its head that looked just like horns. The owl stared down at Henry crossly with its big yellow eyes, and then suddenly it flew away through the woods, going, **"Whoosh! Whoosh! Whoosh!** Whoosh! Whoosh! Whoosh! Whoosh! Whoosh! Whoosh! Whoosh!" until Henry could not see it or hear it any more.

But just then there was another sound inside the tree. And it also went "Pshflelshcrshlopflschlickl!" And then it too went "Ploosh! Ploosh! Ploosh! Ploosh!" right up the trunk of the tree, higher and higher, and Henry's eyes went up the tree with it. And when the ploosh-ploosh-ploosh-plooshes got to the top of the tree another something flopped out of the top of the hollow tree and looked straight down at Henry just like the other one had, also

11

with round stern yellow eyes. It was another big owl. And it stared down crossly at Henry, with its great yellow eyes, and then abruptly it flew away too, going "Whoosh! Whoosh! Whoosh! Whoosh! Whoosh! Whoosh! Whoosh! Whoosh! Whoosh! Whoosh!" until Henry could not see it or hear it any more.

12

Now Henry knew that a family of owls was living inside the hollow tree. He decided that probably the two owls that flew away were the father owl and the mother owl. He also thought there must be some baby owls inside the hollow tree. But they wouldn't be able to go "Pshflelshcrshlopflschlickl!" And they couldn't go "Ploosh! Ploosh! Ploosh! Ploosh!" to the top of the tree and sit and stare down at Henry with big round yellow eyes. And they couldn't go "Whoosh! Whoosh! Whoosh! Whoosh! Whoosh! Whoosh! Whoosh! Whoosh! Whoosh! Whoosh!" and fly off into the distance, out of sight and hearing.

The owls that Henry thought were still inside the tree couldn't do these things because they would surely be tiny baby owls sleeping in their mother's nest. Henry didn't thump their tree any more because he didn't want to frighten the baby owls. He just took his sassafras walking stick and walked off through the woods to look for Indian arrowheads.

Henry didn't find an arrowhead that day, but he did find a beautiful gray celt, or chisel, made of stone by an Indian a long time ago. It was half buried in the yellow clay of a washed-out bank where the river made a sharp turn. Henry climbed down the bank and dug the celt out of the yellow sticky mud. He held it in the water of the river and cleaned all the mud off it very carefully. Then he took the celt home and put it in a special little box in his room upstairs, where he kept all of his precious Indian things.

Chapter 2

The Crow That Played Hide and Seek

Henry never did find his own hollow tree, big enough to live in. But across the years he found the answers to lots of important questions by thumping on things like watermelons and trees.

One day Henry was walking in a neighbor's cow pasture down by the river. He was looking for arrowheads and unusual stones in the places where the cattle and the rain had scuffed all the grass away in the gullies and dips. Once, when he looked up, far out in front of him he saw a big black crow standing in the grass on the edge of the woods. The crow saw Henry too, and it held its head up high. Its eyes were very bright, and it kept its beak straight out and stared right back at Henry, turning its head sideways to look at Henry one eye at a time, the way crows do.

Henry was surprised that the crow didn't fly as he got nearer to it. Crows always fly when they see people. But this crow did not fly. Henry walked straight toward the crow and kept on expecting it to fly. But the crow still did not fly. It only walked away a few steps, and then looked back at Henry again. So Henry walked still closer to the crow, and it walked even farther away from him, and then looked back again.

Now Henry was really surprised that the crow did not fly. So he started running slowly toward the crow. But the crow only started running too. Henry ran faster and faster, and the crow ran faster and faster too, away from Henry. Soon Henry was chasing the crow, running as fast as he could. The crow seemed to be running as fast as it could run too. But Henry could run faster than the crow. Eventually he caught up with the crow, and he reached down and grabbed it. He held the crow in his hands and talked to it.

He had never before caught a crow and held it in his hands. He said, "Nice Crow! Don't worry, Crow! I am not going to hurt you!" And he looked carefully all over the crow. Right away he saw that something was wrong with one of the crow's wings. Its right wing was only partly there. Half of it was gone. Something must have broken the crow's wing, and the broken part had fallen off. The wing was all healed up, and it didn't seem to be sore any more. But part of it was completely gone.

Now Henry knew why the crow could not fly. A bird must have two wings to fly. This crow did not have two wings. It had one wing and a half wing. A bird cannot fly with only one wing and a half. So the crow could

only run to get away from Henry, and that is why Henry was able to run and catch it. But Henry was surprised that the crow didn't even flap its wings in the effort to run faster.

Chickens flap their wings whenever they are running as fast as they can, and the wings seem to help them go faster. Why doesn't a crow with a wing and a half try to run faster by flapping its wings?

Henry carried the crow home with him, holding it carefully in his hands. He knew that, if he left the crow in the woods, sooner or later a fox would come along when the crow was too far away from a tree trying to find something to eat. Then the fox might catch the crow and eat it. A crow is a bird, and foxes eat just about any kinds of birds they can catch. A fox can run much faster than either Henry or the crow. Henry didn't want the crow to be eaten, so he decided to take it home with him.

When Henry got to his own family's farm, he put the crow in a big cage that was actually made for chickens. Then he put in a dish of water, and some corn and other feed used for the chickens. The crow must have been hungry and thirsty, because it immediately ate the food and drank the water. Henry was pleased because the crow did not act as though it was frightened of him at all. He was sure he had saved the crow's life. He thought the crow was probably unhappy because it couldn't be outside flying anywhere it wished. But he also thought it must be a little happy too, because it was safe, because it had all the food and water it needed, and because Henry was being nice to it.

Henry and the crow became good friends. Each day Henry took the crow from the cage and held it in his hands. He petted the crow and talked to it. He said things like, "Nice Crow! You are my friend, and I am your friend! Some day we can play together!" He didn't think the crow understood all of that, and he didn't really know if a crow would know how to play. But he felt good saying things like that to the crow. And he wanted the crow to understand that when he talked to it that meant the crow's world would always be safe and comfortable and pleasant.

One day Henry decided it was time for the crow to walk around in the grass in the barn lot all by itself. So, on that day, when he took the crow from its cage, he didn't just sit and talk to it, and he didn't put it back in the cage right away. Instead he carried the crow into the middle of the barn lot and placed it on the ground in the short grass there. Then he walked far away and watched the crow to see what it would do.

For a long time the crow just stood in the grass without moving. Henry kept on watching it. After a while he wondered why the crow was looking toward the barn and not moving. Henry looked toward the barn too, and saw a big black tom cat coming from the barn. The big black tom cat was walking straight toward the crow. It walked very slowly. It stared right at the crow, lowered its body, and crept along in the grass. Henry knew that the cat wanted to eat the crow. It was creeping toward the crow, step by step.

He was surprised that the crow did not move. It just stared back at the cat with its beak held straight out and its eyes very bright. The cat came closer and closer. Then it crouched down in the grass and was still for a little while. It watched the crow and the crow watched it. Both were very still. Suddenly the cat leaped high up into the air and its body was hurtling straight toward the crow. Just as it was

about to land on the crow's head, with its claws out, the crow went peck, peck, peck, peck, peck, very fast, right on the front of the cat's head. The cat fell over sideways into the grass. It said "Meeeeeoooouw!" really loud, and jumped up out of the grass. Then it ran as fast as it could go all the way back into the barn and disappeared. It was frightened of the crow because it had discovered that the crow was strong, and its beak was very sharp.

Now Henry knew that the crow would be safe around the barn cats. Even the biggest tom cat could not eat the crow because the crow was too good at pecking the cat's head. Henry also decided that the crow's wing must have been hurt a long time ago. During that time the crow had been able to figure out exactly how to behave when it could not fly. Most crows would try to fly from almost every attack, but not this crow. This crow stood its ground before the big, dangerous cat and fought back. It didn't even try to run away.

Henry wondered how many battles the poor crow had to fight before it became brave enough to face its enemies and fight back, and how many times it had tried to fly and failed before it learned not to flap its wings uselessly.

After he realized the crow could take care of itself around the farm, Henry left the crow in the grass and started doing his work near the barn. Sometimes he would look back at the crow to see what it was doing. But the crow didn't do very much except watch everything all around it, including Henry. So Henry climbed up into a wagon parked beside the barn and began to scoop some corn into a window of the barn. While he was doing this he couldn't keep his eyes on the crow all the time.

One time, after he had scooped corn into the barn for a while, Henry stopped and looked up to see what the crow was doing. He had stopped to look before, and the crow was always there. This time the crow was gone. Henry looked and looked, but he could not see the crow anywhere. He couldn't see a single place where the crow might be hidden. But there was no place where it could have walked out of sight either. Where was it? It was gone. Or it seemed to be gone. Where could it possibly be?

Henry just stood still and looked, and looked some more. He kept on looking for a long time but he did not see the crow. Then, while he was still standing in the same place, and just looking, Henry happened to move his head a little to one side. To his surprise he saw a tiny bit of something black behind a skinny post in the fence by the barn. The fence was a long ways from Henry, and it was also a long ways from where the crow had been. And the fence was exactly between Henry and the place where the crow had been. When Henry moved his head a little more to the side, the tiny bit of black seemed to move too, and it disappeared again behind the skinny post. Then Henry moved his head the other way, and this time he saw another tiny bit of black on the other side of the post. This tiny bit of black moved out of sight too, as soon as Henry saw it.

Henry knew the tiny bit of black color was the edge of the crow's shoulder. He understood that the crow was standing directly behind the skinny post. The post was just wide enough that when the crow stood very carefully behind it, with its body in exactly the right position, Henry couldn't see it at all. The crow was hiding from Henry! It seemed to be playing a game of hide-and-seek. Henry moved, and the crow moved. He moved the other way and the crow moved the opposite way. Henry smiled because he immediately liked this funny game of hide-and-seek with the crow.

After that Henry played hide-and-seek with the crow nearly every day. When he looked at the crow it stood still. When he wasn't watching, the crow moved. Sometimes it would hide behind a post, other times behind a box or a bush. When Henry was looking elsewhere, it would slip into the garden and hide under a broccoli leaf. If Henry went over by the broccoli plant, the crow would hurry to the other side, staying under the outspread leaves so that Henry couldn't see it.

After a while Henry realized that hiding might not be just a game for the crow. Maybe this crow had learned to hide or be eaten! Henry wondered if it had learned to hide so well only after it had become unable to fly.

Each day, when their game was finished, Henry put the crow back into the cage where it would have food and water, and be safe from foxes and other animals, bigger than cats, that might eat a crow.

One day Henry stood on the porch steps of the house and held the crow in the palm of one hand. He reached as high as he could. He let the crow step off his hand and stand on the edge of the roof of the house. Then he walked away and turned around to watch the crow and see what it would do. For a long time the crow looked over the edge of the roof and just stared down at the ground. It walked a little this way then a little that way, and looked over the edge of the roof again.

Finally it just jumped right off the roof. This time it tried to use its wing and a half. But it was only able to flutter awkwardly down, turning in a half circle because of the half wing, and land safely in the grass. Now Henry knew that the crow could never fly again, because of the half wing. Except for when it fluttered down from the roof that one time, Henry never again saw it lift its wings or try to flap them. He felt a little sorry for the crow because it couldn't fly, but he was happy that the crow had figured out how to live perfectly well in a way different from its earlier life. He decided the crow had to have a certain kind of smartness that it was using to live its life. He was happy that he was able to help the crow in its effort to live a happy and safe life.

Henry also decided the crow must like him, and its new life, because after a while it didn't try to run away from him when he picked it up to put it back in its cage, with the food and water. And it always played hide-and-seek with him when he let it stay out in the barn lot or the yard. Henry thought the crow must have learned to play hide and seek when it got too far away from a safe tree it could climb, and other dangerous animals like foxes, and hawks, and maybe raccoons, appeared that might try to catch the crow to eat it. But Henry didn't believe the crow thought he was going to hurt it, or eat it. So Henry liked to think that when the crow hid from him it was really just playing a little game with him.

Henry and the smart crow that had only one wing and a half kept on happily playing hide-and-seek together. They remained friends with one another for such a long time that it seemed like forever.

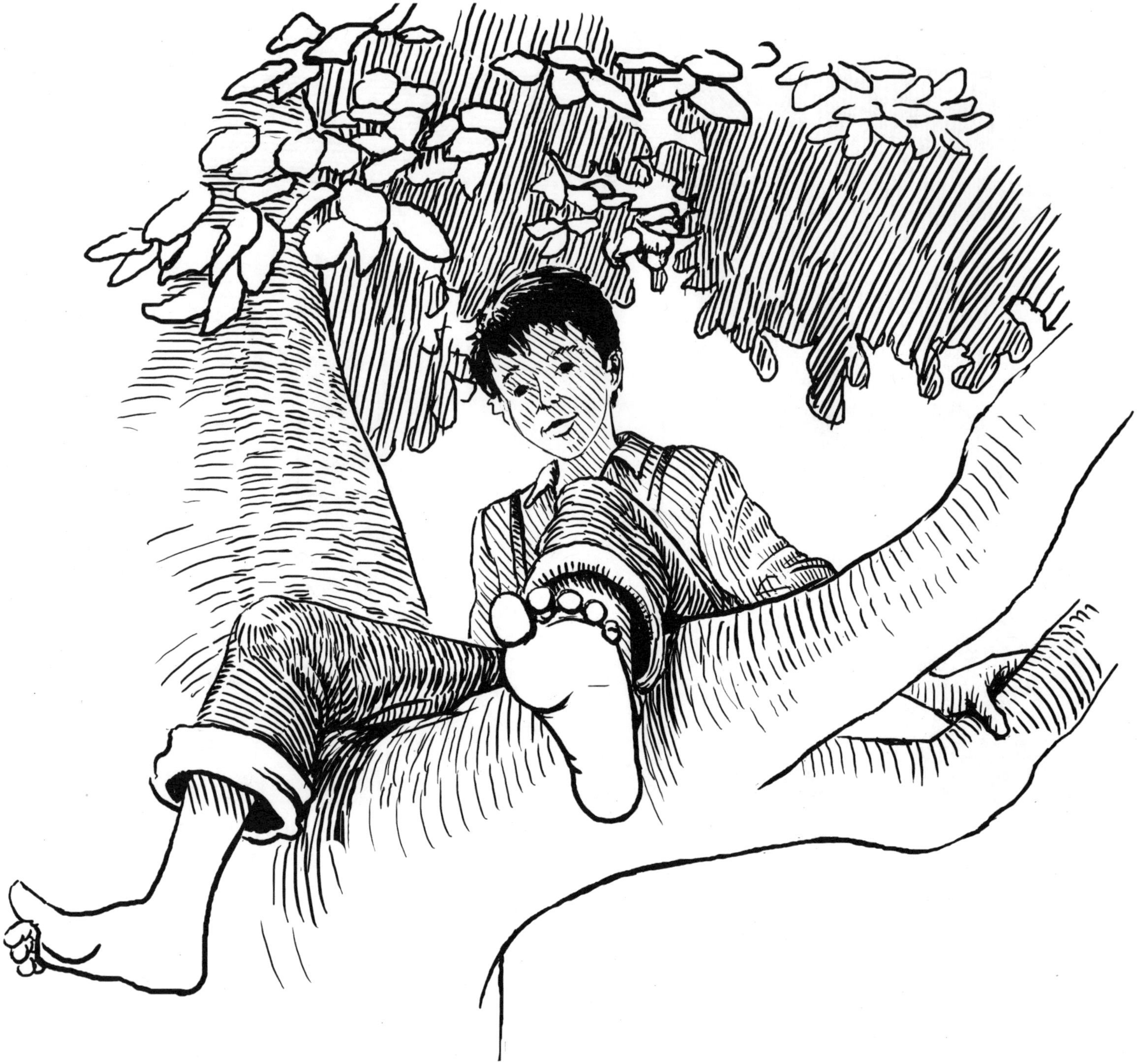

Chapter 3

Going Barefoot

There were few things that Henry liked more than going barefoot. All summer long, all over the farm, and wherever he went, Henry went with bare feet. He didn't wear any shoes, and not any socks either. He just walked barefoot in the grass, on the rocks, in the road, in the barn lot, in the cow poopies, and sometimes on boards with nails in them that hurt. He especially liked to climb trees in his bare feet.

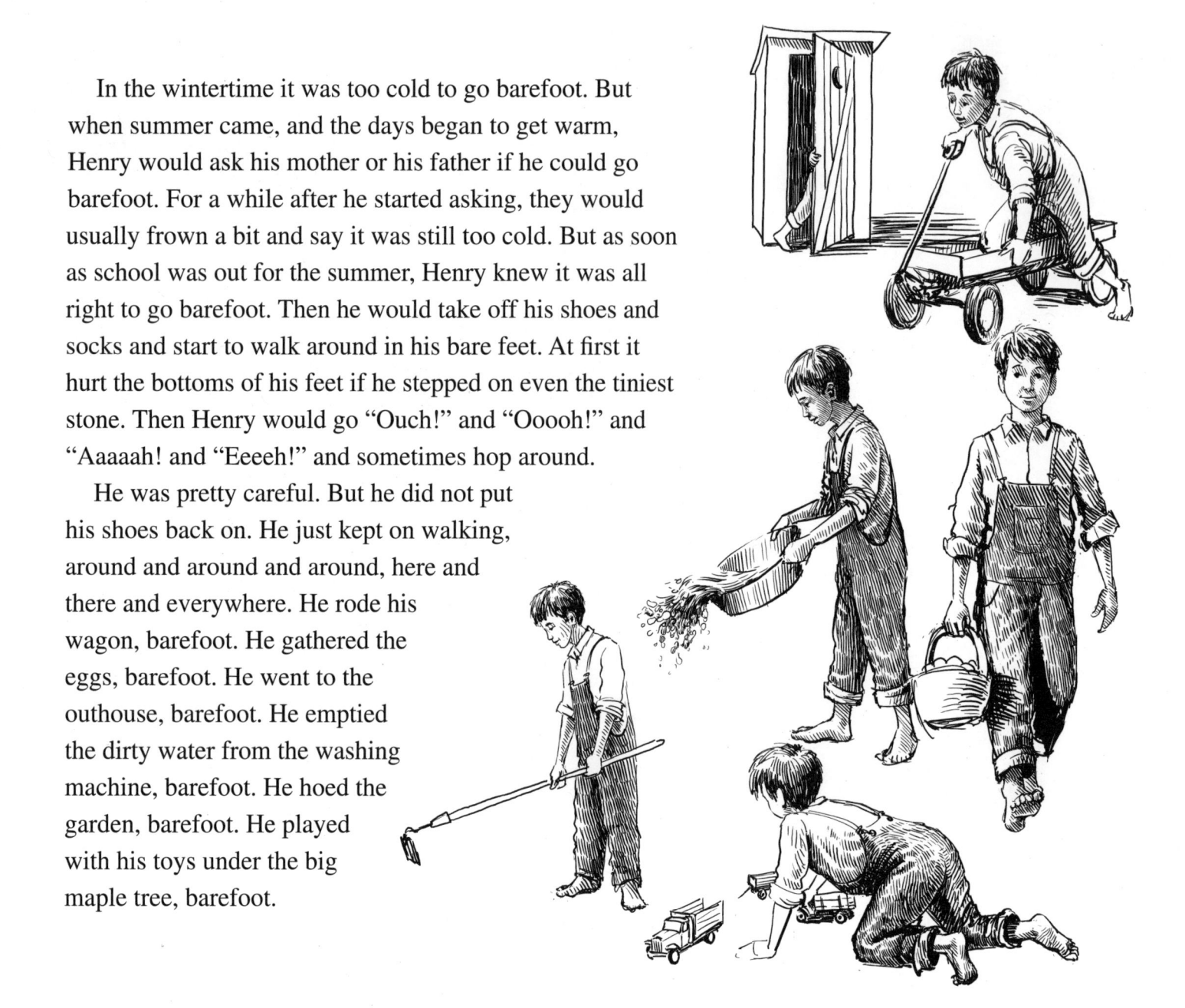

In the wintertime it was too cold to go barefoot. But when summer came, and the days began to get warm, Henry would ask his mother or his father if he could go barefoot. For a while after he started asking, they would usually frown a bit and say it was still too cold. But as soon as school was out for the summer, Henry knew it was all right to go barefoot. Then he would take off his shoes and socks and start to walk around in his bare feet. At first it hurt the bottoms of his feet if he stepped on even the tiniest stone. Then Henry would go "Ouch!" and "Ooooh!" and "Aaaaah! and "Eeeeh!" and sometimes hop around.

He was pretty careful. But he did not put his shoes back on. He just kept on walking, around and around and around, here and there and everywhere. He rode his wagon, barefoot. He gathered the eggs, barefoot. He went to the outhouse, barefoot. He emptied the dirty water from the washing machine, barefoot. He hoed the garden, barefoot. He played with his toys under the big maple tree, barefoot.

He practiced the piano, barefoot. He swung on the swing his grandfather had built for him and his sister, barefoot, and he even jumped out of the swing in his bare feet. That hurt! But he kept right on doing it. He got the cattle up from the pasture to be milked, barefoot; and he took them back after milking, barefoot. And very slowly his feet became tough. They became so tough he could walk anywhere and it didn't hurt a bit. Well, almost anywhere.

Sometimes he stepped on a sharp stone or a piece of broken glass or a nail in a board. Then he would go "Oooooh!" Sometimes he would even cry a little, and his foot might bleed. And it might hurt for a very long time. Then he would walk funny too.

33

If the nail went in the back of his foot he walked only on the front of the hurt foot.

If the nail went in the front of the foot he walked only on the back of the hurt foot, and he had to twist the foot sideways just before he picked it up to take the next step.

If the nail went in the outside of the foot he walked only on the inside of the hurt foot.

If the nail went in the inside of the foot he walked only on the outside of the foot. But that was especially hard to do.

If the nail went in the middle of the foot he hopped on one foot because it hurt too much to put the sore foot down.

Sometimes Henry stubbed his bare big toe against a rock or a board, and that would also hurt very badly. And he would yell and scream and dance and hop around, and sometimes he would hold his leg up in the air and yell and hop around some more. Everybody laughed at Henry when he walked all of the funny different ways after stepping on a nail, or when he stubbed his toe and danced around. They would say, "Why don't you just wear your shoes, Henry?" But Henry didn't say anything. He just kept right on going barefoot because that's what he liked to do.

Sometimes Henry didn't watch where he was going, and stepped in a cow poopie, and then his foot would be a strange greenish color for a while, and also have a very special smell. And his mother would say, "Henry! Wash your feet before you come in this house!"

Sometimes, on very cold mornings, Henry and his brother would step into a fresh cow poopie on purpose and just stand there in the warm cow poopie until their feet warmed up. The warm cow poopies felt really good, and Henry and his brother knew that after they had gone barefoot all day long there wouldn't be a trace of cow poopie on their bare feet.

Sometimes at night Henry forgot and started to go to bed without washing his feet, and then his mother would call out to him in a very definite way that Henry understood perfectly. She would say, "Wash those dirty feet before you go to bed!"

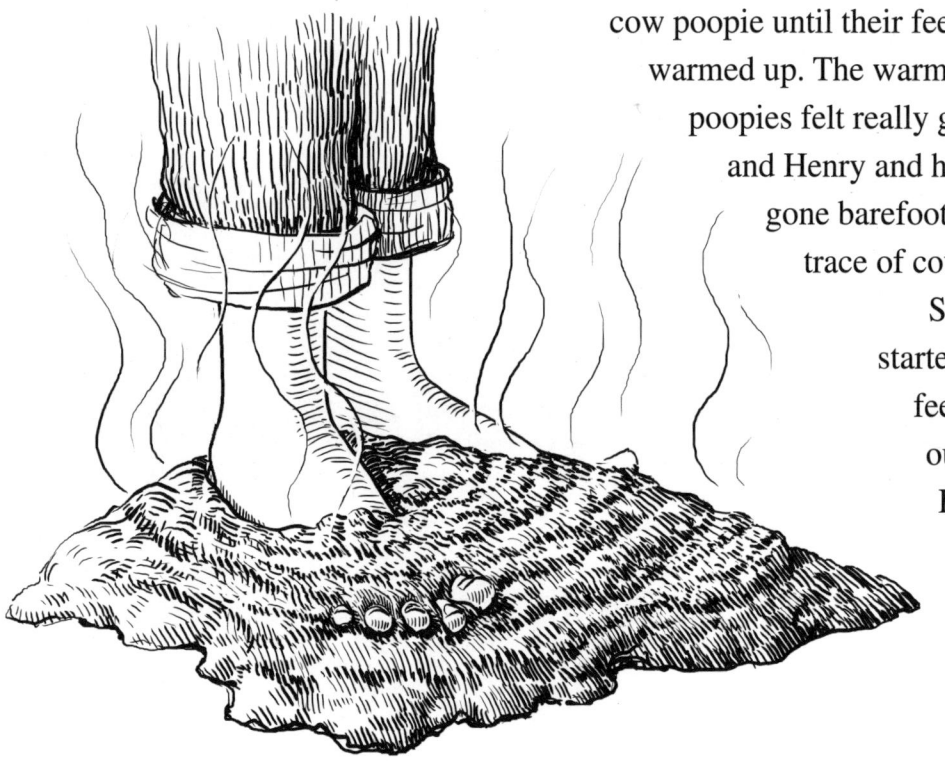

Once in a while Henry decided his feet weren't too dirty after all, and he would try to sneak off to bed without washing them. But his mother had sharp eyes, and she always saw his dirty feet and cried out. And then Henry would look funny and duck his head and go outside to wash his dirty feet at the water pump.

Henry liked to go barefoot because it was easier to get out of bed in the morning and go outdoors if he didn't have to put on shoes and socks. And he also liked the way the grass and the ground felt under his bare feet. He could feel all kinds of things under the bottoms of his feet. Sometimes the ground was cool and sometimes it was warm. Sometimes the rocks or the road were too hot, and then he would walk funny in a different way. He would hop on one foot and then on the other foot and say "Ooh! Ooh! Ooh!" and everyone would laugh at him again. But Henry didn't care. He liked going barefoot.

In the springtime when his father plowed his fields, Henry walked along behind the plow with his bare feet in the cool furrow. He liked looking at the soil as the plow turned it up and over and left the damp furrow for his bare feet to pad along. He could hear the kildeers and the crows, and the jingle of the harness of the horses that pulled his father's plow. And he just walked and walked, behind the plow, with his head down, watching the new black soil that the plow turned up, and listening to the birds, and smiling to himself.

One day Henry was walking behind his father's plow with his head down and his bare feet feeling the new damp soil, just watching the good clean soil turning up and over, and smelling the freshness of the soil, and hearing all the birds singing and the harness jingling and the horses puffing and blowing as they trudged along pulling the plow. Suddenly, he saw something sticking in the dirt, right in the side of the new furrow he was walking in. It was bright red, and at first it looked like just a sharp stone. But it was a very bright red. And Henry could see that it was a beautifully shaped Indian spearhead.

He bent down and picked it up and looked at it, and then he ran as fast as he could after the plow and the horses and showed it to his father. His father turned the spearhead over and over and told him a story about how the Indians had lived a long time ago in the same place where Henry was walking in the plow's furrow. A thousand years, or maybe more, before Henry found the spearhead, some Indian had lost it there. Henry's father thought the Indian must have been a very important chief to have such a beautiful red stone head on his spear.

After that Henry liked to imagine that he was an Indian too, whenever he walked down the hill by the creek and hopped from stone to stone with his bare feet, or when he rode his horse, Major, down through the pasture. He imagined that he was a famous Indian warrior who rode his horse with his bare feet hanging down on either side, and carried a long spear with a great red stone spearhead on the end of it.

Sometimes, in the summer, Henry took piano lessons. Then he would have to ride with his mother to a nearby town and go to a certain lady's house and play her piano for an hour or so. Whenever he went for his piano lesson, his mother made him put on his shoes. Sometimes she would drive down the road until she saw Henry playing in a field or along a fence row. Then she would toot the car horn and call out to him. Henry would run over to the car and jump in the back seat. There he would see those horrid shoes waiting for him.

And his mother always said promptly, "Put on your shoes, because we're going to be there pretty soon." And Henry would pick up the shoes and the socks, and sigh, and slowly put them on. His feet burned and burned when he put on the shoes and socks, and it seemed to Henry that the feet wanted to be bare again. As soon as the lesson was over, he took off his shoes and socks and told his poor feet how sorry he was to make them burn and burn like that. The feet never said "Thank you!" But Henry felt good anyway, about being nice to them.

When school started in the fall Henry had to put his shoes back on for the whole winter. The school teacher didn't allow students to come to school with bare feet. Henry didn't like to put his shoes on, because he knew that for a week or two his feet would feel as though they were just burning up. They didn't like having shoes and socks covering them up after being nice and bare all summer.

But Henry knew that spring would come again, and then off would come those shoes and socks, and once more he would be able to feel the grass and the soil and the rocks, and climb up the trees and feel the rough bark with his feet.

And he also knew that sooner or later he would step on something sharp and hop around and walk funny again, and everyone would laugh at him. But he would not say a thing. He would just go barefoot, all over the farm, and look for more Indian arrowheads.